The Cantelonian System of Hatching Eggs and Rearing Poultry

by Wm. Jas. Cantelo

with an introduction by Jackson Chambers

This work contains material that was originally published in 1848.

This publication is within the Public Domain.

This edition is reprinted for educational purposes and in accordance with all applicable Federal Laws.

Introduction Copyright 2017 by Jackson Chambers

Self Reliance Books

Get more historic titles on animal and stock breeding, gardening and old fashioned skills by visiting us at:

http://selfreliancebooks.blogspot.com/

Introduction

I am pleased to present yet another title on Poultry.

The work is in the Public Domain and is re-printed here in accordance with Federal Laws.

As with all reprinted books of this age that are intended to perfectly reproduce the original edition, considerable pains and effort had to be undertaken to correct fading and sometimes outright damage to existing proofs of this title. At times, this task is quite monumental, requiring an almost total "rebuilding" of some pages from digital proofs of multiple copies. Despite this, imperfections still sometimes exist in the final proof and may detract from the visual appearance of the text.

I hope you enjoy reading this book as much as I enjoyed making it available to readers again.

Jackson Chambers

PREFACE TO THE SECOND EDITION.

Since I first issued my "Practical Exposition of Artificial Incubation," I have had many temptations to increase its size, by replies to arguments, and observations advanced by individuals with whom I have conversed, and who have honoured me with their correspondence; but as my talent does not lie in the way of writing a book, I shall still confine myself to the strict matter connected with my discovery and invention, for the purpose of inducing individuals, or companies, to carry extensively into practice my system of hatching eggs and rearing poultry, as an article of food for the community. My process in no degree alters the nature or character of the fowls. It only facilitates their production, so that the most approved methods, recommended by other authors, for the feeding and management of poultry, need not, by my plan, be superseded. With regard to the commercial, agricultural, and profitable department of the business, I have added, in an appendix, a list of prices; some remarks, and an estimate of the probable profits connected with the business.

Regarding my scientific discovery, I now quote the advertisement which appeared in the *Times* and *Morning Post*, of the 8th January, 1848.

"Model Poultry Farm, Chiswick, Jan. 7th, 1848.

"I hereby give notice to the public, and to the scientific world in particular, that some years ago I discovered the blood-heat of the feathered tribe to be 106 degrees Fahrenheit, instead of 98 degrees, that of the human race, as hitherto asserted and believed, and I challenge the world to produce evidence to the contrary. The design of the Great Creator in such an arrangement, and other particulars connected with the subject, I will endeavour to prove, and also, that top-contact heat, and two temperatures applied, is the only true principle of hatching eggs and rearing poultry.

"WM. JAS. CANTELO."

In elucidation of the subject, I now quote a letter which I had the honor of reading and presenting to H.R.H. Prince Albert, in Windsor Castle, on the 4th January, 1848, when I had the distinguished honor of exhibiting my invention to the whole of the Royal Family and Court.

"My Prince,—

"It has been asserted and believed, that the blood heat of the feathered tribe is the same as that of the human race, viz., 98° Farenheit, and, therefore, ovens have been used at that degree, which caused the eggs to evaporate over the *whole* surface, thus destroying in almost every instance the principle of vitality. This is a fatal error, as I shall plainly prove. Were this the degree of heat required, the atmosphere alone would in many countries supersede incubation, and hatch the eggs without the assistance of the parent bird, or "top-contact heat," the natural principle of hatching. This is evident from the fact, that the thermometer occasionally rises in many parts of the world to 98 and even 100 deg., and should the atmosphere continue at this point during eighteen hours, vivification would commence; but it would be arrested if the degree of heat should fall below this point; as, also, animation would infallibly be destroyed, should the atmosphere remain at the same degree of heat, in consequence of the undue evaporation going on over the whole surface of the eggs. My great experience has led me to discover, that the blood heat of the feathered tribe is 106° Fahrenheit; a higher degree than is ever known in any country to continue long enough to vivify the eggs; and that during incubation, this heat is communicated to a very small surface of the eggs, and there maintained by "contact at the top." This explains the "two temperatures;" viz., that caused by the contact of the breast of the bird during incubation; the other, the surrounding atmosphere, according to climate, season, or circumstances, varying from freezing to 100°.

"WM. JAS. CANTELO.

"*London, March*, 1848."

INTRODUCTION.

In putting this little treatise before the public, the author has endeavoured to place facts in plain, simple, and short terms, as well for the benefit of the farmer, as of those who may wish to practise Hydro-Incubation. Every idea will be found expressed in as few words as possible, divested of all technicality; and the reader may rely on all the information being founded on long experience and close observation.

These pages might have been lengthened by philosophical remarks, extracts from various works, and reasonings *pro* and *con;* but in so doing, their utility would have been lessened, since few persons have either time or inclination to sift facts from useless verbiage.

In cases where the author's experience does not coincide with the observations of other writers, he has added remarks, but he trusts to the great arbiter —Nature, for the elucidation of his assertions. He has not thought it necessary to refer in any manner to the works of others, except in a general view; and it is hoped that nothing herein contained will be supposed to point at any in particular.

GENERAL REMARKS AND OBSERVATIONS.

I SHALL commence this subject by asserting as an indisputable fact (how much soever it may be opposed to the generally received opinion) that *No means have heretofore been discovered for Artificial INCUBATION, in the slightest degree resembling the operation of Nature.*

Although the Egyptian method of *hatching* has proved the most successful of any hitherto attempted, still it does not owe its continuance to the correctness of its principle, but to isolated facts, some of which are as follows:—

1st. In all hot countries the various kinds of animal food are comparatively scarce; consequently, the breeding of poultry is a matter of paramount importance.

2nd. The fowls lay more eggs, and are less prone to incubate than in our more northern climates.

3rd. The chickens are more easily reared without the maternal assistance, owing to the equable temperature of the climate, and the care that can be afforded them at little cost.

Thus we see that with only partial success what *can* be effected helps to fill the void, and pays in price what is wanting in quantity.

It is true that artificial *hatching* has been extensively tried, and publicly exhibited; but the failure of these experiments testifies that the means employed have never been adequate to the end in view. And yet these pretenders have all drawn back from admitting their failure in a process universally supposed to be so simple as only to require an even blood-heat during twenty-one days; all have attributed their ill success to the want of a regular temperature, or proper degrees at proper times; *all have overlooked the meaning of the word* " INCUBATE," " *to sit upon,*" and the necessity of carrying out in their experiments the principle involved in that expression.

Many have been led to continue their labour for years, from the fact, that, although the principal injury takes place during the first days of the process, its fatal effects only appear towards the time when the chicken should release itself from the shell. Others have sooner abandoned the attempt; but ask them why, and you will be told—other business, the difficulty of procuring eggs, or some excuse equally remote from the real cause, viz., they did not find it so simple as it appears.

There have been millions spent in these fruitless attempts, which might have been saved, had Reaumer, and Bonnemain, and others, told the truth instead of pretending to a real success which these pages will *prove they never had:* nay, *it would have been better had they even remained silent on the subject*, instead of enticing others, by delusive representations, to follow blindly in their steps. Many have done so to their ruin; and this was the cause of the loss of several thousand pounds to the inventor of the present process, although eventually leading to the desired end.

The plan, heretofore pursued, has been to place the eggs in a room, or oven, heated to 98° or 100° Fahrenheit. According to the opinions of different advisers, the heat should be regular throughout the time; or raised, or lowered gradually, to or from this point,—all recommend placing water in the ovens, in order that the evaporation may serve in place of the natural moisture of the fowl.

All this looks very natural at first sight, but is easily proved to be fallacious. In the first place, there is little or no variation in the blood-heat of the fowl; it is at all time about 106 degrees, being several degrees higher than the blood-heat of the human race. In the second place, the fowl throws out very little perspiration or moisture; and if she did, it would neither enter the warm egg, soften the shell, nor prevent the egg from evaporation. But it is hardly necessary to disprove, piece by piece, all the fallacies which have been put forth on this subject; as a clear exposition of this natural process effects the same object as a whole.

Nature works with consummate skill, and all her operations are perfect; we, therefore, must not try to improve her ways, but follow implicitly in her steps, in all our attempts to imitate her wonderful workings. She has ordained the germ of the egg (*so long as it is kept in a horizontal position*) to float uppermost within and against the shell, in order that it may meet the genial warmth of the breast of the fowl. We must, therefore, in incubation apply warmth to that part only, and of the degree determined by nature.

A fowl of any kind prefers to incubate upon the ground. Nature having supplied the egg with only a limited quantity of moisture, has thus arranged to prevent evaporation from a large surface, as the egg is only warm at the part in contact with the fowl, until the blood-vessels forcing nourishment for the embryo, have surrounded the inner surface of the shell; when, thus, the whole egg becomes gradually warm, and eventually of an equal temperature, by means of the circulation of the blood through these vessels.

We must, in a word, apply the same degree of heat as Nature, and in the same manner—"by top contact,"—and, like her, allow the inferior portion of the egg to remain cool, until warmed by the inward circulation of the blood.

The difference between "top contact heat," and that received from radiation, as applied to hatching, is this:—by radiation, or oven heat, the eggs will be hours arriving at the desired temperature, not only when first put to hatch, but at any time afterwards, when they may have been allowed to get cool. The eggs will, of course, heat alike over their whole surface, and, consequently, evaporate equally from every part. On the contrary, heat applied in "top contact," penetrates almost instantly, and revivifies the germ; and although a much higher temperature is used in this case, in imitation of Nature—that is, 106, instead of 98 degrees— still, inasmuch as but a small surface is thus heated, the loss of moisture is much less than by a radiating heat.

The fowl leaves her nest every day, in search of food, for twenty or thirty minutes;* this must be imitated also, as the temporary loss of heat has the effect of causing the contents of the egg to diminish in bulk, and the vacuum is filled by a fresh supply, drawn in, for the nourisment of the germ.

The eggs must be moved three times a day—morning, noon, and night—which prevents the adhesion of any part of the fluid to the shell, and also gives the small blood-vessels a better opportunity to spread around the surface of the egg. This is effected by Nature; when the fowl leaves her nest, or returns to it, she naturally disturbs the eggs, and also, from any change she may make in her position while upon her nest.

THE HYDRO INCUBATOR.

The form or method, considered by the inventor as best calculated for the application of "top contact heat" to eggs during incubation, is that of a current of warm water flowing over an impermeable or waterproof cloth, beneath which the eggs are placed. This

---

* From experiments which I have made, I am satisfied that should a hen leave her nest for the space of twelve hours and upwards, the hatch would not be injured. It may appear still more extraordinary, that the vitality of an egg (provided it has not been sat upon) is not liable to be destroyed by exposure to cold, however intense, unless the shell be burst.

is effected, on a large scale, by pumps, and, in a small apparatus, by the law of gravitation causing the particles of warm water to rise, and those that have become partially cooled to fall. A tank of water is kept continually at a temperature of 109 deg. from the surface of which it will naturally flow over the waterproof cloth, a return-pipe being so placed as to connect the outer end of the cloth with the bottom of the tank. The eggs are placed in drawers having open-work or perforated bottoms, and they are laid on a piece of thin woollen cloth. The drawers are placed beneath the Incubator, and raised so that the eggs come in contact with the waterproof cloth, but so as to allow a space between the sides of the drawers and the incubating cloth. These sides being lower than the top of the eggs, allows the air to circulate around them, as it rises through the bottom, and passes out over the edges of the drawers.

These particulars, not very easily conveyed in a limited description, will be duly appreciated on inspecting an Incubator in operation.

PROCESS OF INCUBATION.

The eggs intended for incubation, having been inspected as hereafter described (which can only be dispensed with in case of perfect confidence in the source of supply), are placed under the incubator.

The eggs should be gently moved three times in the twenty-four hours, and once a day they ought to be slightly damped with a soft sponge, on the top only, as they lie in the tray. About mid-day, daily, they must be taken out and allowed to cool for twenty or thirty minutes, thus imitating the action of the hen leaving her nest to feed, and allowing a fresh supply of vital air to pass into the egg as it cools.

After three days of incubation is the surest time for inspection, in order to take out those which have no germ; although it may be done previously; as, after eighteen hours of incubation, you will observe, at the top of the egg, a round spot or shadow, which is the beginning of the germ; but not so certainly as at the expiration of three days. They must be again inspected after the tenth day; this second inspection will prevent those having only an imperfect germ from becoming addled, and injurious to the good eggs. There may still be an occasional bad egg after this time, but it will be easily discovered by practice, from a ringing or hard sound, when moved with the others; while the good eggs sound very dull, as if cracked. A suspected egg should at any time be examined, and withdrawn.

The hatch should begin pecking at the expiration of nineteen days and a half; thus, supposing a number of eggs to be put to incubate on a Thursday at 5 P. M., on the Wednesday morning previous to the expiration of three weeks, I should expect many to have pecked, and some even to begin to come out. Those which have not hatched of their own accord, on the Thursday morning three weeks, may be reckoned (provided the heat has been kept up to the right point) as good for nothing, even if taken out of the shell; that is to say, those which are last are worth least.

THE ARTIFICIAL OR HYDRO MOTHER.

This consists of a number of warm pipes about an inch and a quarter in diameter, and about the same distance apart, resting on supports about five inches from the floor. Beneath these pipes is a sliding board, which is always at such a height as to allow the backs of the chickens to touch the pipes, and which is gradually lowered as they increase in size. This board is removed and cleaned every day, or replaced by another, which had served the day before, and had been cleaned and aired during the twenty-four hours preceding. Above the pipes (about an inch) is another board, similar to that below, from which depends a curtain, in front of the *mother*. This board serves the double purpose of economising the warmth, and preventing the chickens from dirtying each other, as they are very fond of jumping up on the *mother*, if not prevented.

The pipes above described proceed from the small tank of warm water, the heat being kept at about 109 degrees. The young chickens having been once placed beneath this mother, will only leave it to eat, drink, and exercise, and will return to it of their own accord.

I have had great success in rearing turkeys and guinea fowls by this artificial mother, and will undertake to rear a greater proportion by this means (turkeys especially) than can be done by the parent fowl. This may, at first, appear to be a startling assertion; but let all things be considered before judgment is passed. 1st. The fowl has invariably some vermin, and often a great many lice. These are spread through the brood which she fosters, much to their detriment and discomfort. 2nd. The fowl often tramples on the chickens—this always injures, and sometimes destroys them. 3rd. If the chickens of one brood go near the mother of another brood, they are generally pecked or killed. 4th. If the brood is following the hen, it is often over-fatigued, and fewer come home than she took with her. All these points shew a great advantage in favour of the artificial mother.

FEEDING.

Relative to the feeding of chickens a great mystery is made; some recommend bread crumbs, others, bread or toast sopped in wine or ale, and a great many other recipes; but I would inquire whether, if these were necessary, nature would not supply them? On the other hand, what *does* she supply? Seeds, grain, grass, and worms;—give these, or cracked grain, grass, and worms, or a little chopped meat, and the chickens will thrive.

CHOICE, AND CARRIAGE OF EGGS.

The egg, when fresh laid, if inspected against the light, will exhibit (generally at the butt) a small void, about the size of a fourpenny-piece, which increases in size from day to day, by the evaporation of the moisture of the egg; thus giving the means of judging amongst a number which are the freshest, and consequently best adapted for incubation. It is not meant to be understood that no eggs kept a considerable time will hatch; but merely that the chance is much against them. I have, however, hatched a portion of some eggs kept in a temperate atmosphere upwards of two months, they having been turned regularly every day. All eggs of irregular form, having two yolks, or any part of the shell very thin or scaly, or with any crack or flaw, should be rejected. Should, however, any valuable or rare egg have a defect in the shell, it may be worth while to gum a piece of paper over the part effected, as it is through the extra-evaporation that it would otherwise fail.

Much has been said relative to the injurious effects of the transport of eggs for incubation, and it has even been asserted that carriage by water is injurious. I do not say that an egg, purposely shaken with violence, will produce a chicken. This I have never tried; but I can say that they will hatch very well after an ordinary carriage of thirty or forty miles over country roads, provided they have been well packed. I have hatched many fine chickens from eggs which had travelled by rail one hundred miles, and by carrier sixty, having been bought previously in the market of a country town.*

Eggs are generally packed in straw, bran, or, chaff; there is, however, a packing much superior to these, which I have adopted with success, viz., Oats. This is, of all others, the most economical packing for eggs; for whilst the packer supplies the other at his own cost, he reaps several advantages from using oats. He charges the current price for his oats; he will have no broken eggs (a great item); the eggs are packed in smaller compass, and unpacked with a better appearance; they require much less time to pack, as the oats are thrown on in alternate layers with the eggs, fill up all interstices, and the two together form almost a solid body.

---

* However to insure success I would in all cases advise those working my Incubators to keep their own producers, at the rate of one cock to four hens, and use only their own new laid eggs.

DEFORMITIES.

Stale eggs often produce ill-formed feet or legs, and the same effect is produced by oven-hatching, and even by the present process occasionally, when the water is kept at much too low a temperature; but, with a proper heat and fair eggs, a deformity of the chicken will scarcely ever be found under the Cantelonian System.

In all cases of deformity, it is most economical and humane to destroy the chicken. If a *cross-bill*, it always grows worse, and will finish by not being able to eat at all, and a *stiff-leg* is pulled about, and made miserable by the other chickens; and, inasmuch as a deformed chicken would not have left the nest of the mother, it is not worth while to attempt to do better artificially.*

One thing is to be particularly observed, in opposition to what is often written:—Never attempt to free a chicken from the shell, unless the cause of its detention is very evidently an accidental circumstance, which you may know by its loud cries, sometimes caused by the feathers sticking to the shell; but when a chicken is nearly disengaged, or making very violent efforts, there is no danger in pulling open the shell, though the least abrasure of the veins covering the inside of the shell before the blood is taken up by the chicken is always detrimental, and generally fatal. In case, however, of the chicken pecking towards the small end, instead of the butt (which sometimes happens), as soon as it begins to cut round the shell, a piece may be removed, in order to give a little more room for the exit. As the chickens come out, they

---

* I have hatched a duck with three legs, that is, an imperfect and extraordinary one, proceeding from below the root of the tail. This lived and did well, as it had two good legs to stand upon; but the third one was often pulled at by the others.

are gathered in a warm place over the incubator, or tank, in order, when dry, to be placed under the *mother*.

PRODUCE.

I will at any time undertake to produce more good chickens from a hundred eggs, than can be shown to be hatched by fowls from another hundred out of the same basket, provided always that the fowls shall have no more than ordinary attention bestowed on them, and provided that this is done by disinterested persons. I shall gain by the accidental circumstances attendant on the incubation of fowls.

General Directions.

CLEANLINESS is of great importance, and in this respect you cannot be too careful, particularly where a great many poultry are to be reared. Their place should be cleaned, and well ventilated, every day. Dirty water is very hurtful to chickens of any age, and will alone be sufficient to breed distemper, and cause them to die without any apparent cause.

At six weeks old, the chickens should be removed from the mother, and placed to roost on small perches, three feet and a half from the ground, in a warm place, and every evening, when they go in, they must be put up to roost, as you have no fowl to entice them; in a few evenings they will go up of their own accord.

Too great crowding of the chickens must be avoided at all times, as this of itself will create disease. Should any illness appear, such as sneezing, or watery or sore eyes, those affected must be picked out with the greatest care, and killed. This will root out the evil; whereas, the time spent in curing a chicken does not pay, and suffers the disease to spread; therefore, I say, nip it in the bud.

There has been fully as much nonsense written on the diseases of poultry, as upon artificial *hatching*, but I could never discover the diseases, nor the pretended causes. Disease is, for the most part, engendered by bad feeding, bad water, bad air, or want of cleanliness, and so on; and the only remedy is to root out the cause.

In rearing chickens with the artificial mother, we must not run into any extreme. Inasmuch as the fowl often gives the chickens too much labour in following her, they must also get too little exercise when she is cooped up. With the artificial mother, it is better to entice them gradually to feed at the further end of the enclosure. This will cause them to spread, and take salutary exercise; but when very young, or during bad whether, they must be fed indoors.

It is good for the chickens to be well fed, but they must not be surfeited; they are naturally very ravenous, and even when full of their ordinary food, will cram themselves to repletion with something more enticing; for which reason I much prefer giving them their mite of meat, or other delicacy, previous to the ordinary meal, or when they are hungry. It is well to let them be hungry once a day.

In speaking of chickens, I wish always to be understood to include turkeys, guinea-chicks, pheasants, partridges, and all young poultry; as they fare very well alike; turkeys are very fond of vegetables of the onion kind, which they eat ravenously, and this food is very good for them.

The different kinds of chickens should be kept separate, or at least separated when a few days old, unless in case of necessity. If turkeys are kept with chickens, it must be with those of a less age, as a turkey chicken is very slow and stupid in comparison.

The ground and situation best adapted for poultry, is a sandy, loose earth, not too elevated, but dry and sheltered.

APPENDIX.

It may be true that poultry have been brought out by artificial means, but the produce has been so inadequate to the expense, that all inventors have abandoned their enterprise, so far as remuneration went. The non-existence, in the United Kingdom, of a system in operation for the purpose of gain, is surely conclusive evidence that all past inventions have been of no practical value. It could not fail to be otherwise, for all sought to maintain one uniform temperature all around the eggs, which more frequently baked the chickens in the shell than hatched them. It would be useless hatching poultry unless they could be well and cheaply bought up as food for the community. The *Cantelonian System* does this; and, by means of it, poultry have been hatched and reared to a great extent.

Allowing a hen two broods, of thirteen eggs each, a year, she will raise only, on the average, eight chickens out of the thirteen. The patent process produces eighteen times a year, on the average seventy-five birds from one hundred eggs, and thousands of them at a hatch, or as many as the incubators may be made to accommodate, and it is capable of being fitted up to produce millions every day.

The plan of production and feeding is so rapid that, at the expiry of ninety-six days, the poultry is fit for market. It is only needful to reserve a few laying stock until they are eight or nine months old; these, then, go to the poulterer, and are replaced with others as they come forward; so that the feeding of old hens is quite out of the question. Can it fail to be obvious the difference of eighteen broods a year instead of ten —hundreds instead of ten at a hatch—and marketing poultry in thirteen or fourteen weeks, instead of one, two, or three years, as is the present practice? The principle applies to ducks, turkies, game, &c., and there is not a nest of pheasants, or partridge's eggs, that might be disturbed by the mowers or other cause, that could not be hatched fully out of the shell, and the birds set upon wing.

Poultry thrive best on a variety of food. All animal offal, liquid or solid, may be used to feed the creatures. Graves, and the refuse of the sugar-refiner, the brewer, distiller, and baker; overplus fish, and broken meat; the outer leaves of vegetables, even some weeds, with tail corn and damaged grain, Indian corn, rice, and the kernel of the cocoa-nut, which last will fatten and finish poultry with a most delicious flavour.

By a small reduction of price, surely much more poultry might be sold, were it produced. Out of 480,000 farms, it is estimated that they do not send to market more than 9 or 10 million head of poultry a year, to supply the whole population of the United Kingdom, shipping and all, which is not one third of a fowl to each person once a year. Were every one to have a fowl as part food, once a month, it would take 326 millions more of fowls than are at present produced, which would require 893 Incubators, giving out 1000 fowls a day, or 250,140 trays of 100 eggs each, to supply the deficiency. Many children and poor old people might be kept out of the workhouses, and engaged in this business. So cleanly and free from smell are the Incubators that they may be put in a parlor, are best accommodated in the basement floor of a house, and may be most efficiently accommodated in a conservatory.

The business is so simple that any one of ordinary capacity may manage it, by following the printed rules sent out with the Incubators; and many whose health is suffering by their confined occupations in towns, might safely resort to a cottage and a garden in the country, to rear poultry for the support of their families, and the production of wholesome and delicious food for the community.

The profits of the business are large, and as all is clear and above ground, embarking in it can hardly be reckoned as a speculation; for there will ever be a demand for food, and even vegetables cannot be produced with greater rapidity than poultry by the *Cantelonian System of Hydro-Incubation.*

Land of the poorest description, and otherwise almost valueless, is suitable for the purpose of a poultry farm, provided it has a dry bottom or is well drained; and the deposit of the fowls would enrich it, to produce good crops of corn, were their walk occasionally changed. The guano left in the roosting houses would go a great way to pay expenses, and even feathers come to something.

Much misconception on the feeding of poultry prevails. It is said to be more costly than feeding beasts; on the contrary, to feed up an ox to 1,200 pounds weight, usually takes five years—to produce the same weight of poultry can be accomplished in three months, at less than half the cost. Thus the return would be quicker; and capital employed in the Cantelonian poultry business may be made to do wonders.

To ascertain the cost of a fowl, I have made many experiments and trials; and, without the run of a farm-yard, I reckon 8d. will amply pay for the food of a fowl up to three months old. One day, namely, on the 7th Oct., 1847, when I had on hand 1270 chickens, of all ages, from a day to ninety-six days old, which were hatched out from a 4-tray incubator which I had in operation, I gave them:—

	s.	d.	s.	d.	
4 quarts of barley	0	8	or 41	8	a qr.
2 ,, wheat	0	6	or 66	8	,,
1½ pint of grits	0	4			
Two-thirds of a bushel of potatoes	0	8	small sorts		
20 lb. India corn meal	2	1	or 20s. a barrel		
30 lb. barley meal	3	0	or 19s. a bag		
Carried forward	7	3			

	s.	d.
Brought forward .	7	3
12 lb. graves . . .	3	0 or 3d. a lb.
Cabbage	0	6
Suet for those in coops	0	4
	11	1

Multiply 11s. 1d. by 96 days, and divide by 1270, and the cost of a fowl, 96 days old, will be found 10d. But as the above were famine and retail prices, it is not too much to state 8d. as the cost of feed of a fowl at that age, and they might be produced for even less.

Estimate of the Cost and Expenses of Working, and Profits, of a 1-Tray Patent Hydro-Incubator, for One Year.

	£	s.	d.
Cost of the incubator	21	0	0
1800 eggs for 18 broods a year, at 1d. .	7	10	0
75 chickens from every 100 eggs, gives 1350 chickens in the year, and the cost of their food 8d. each	45	0	0
Charcoal for the year	4	10	0
	78	0	0

	£	s.	d.
Fowls are seldom lower than 5s. a couple in London, say 2s. each for 1350	135	0	0
Value of incubator . . .	19	0	0
	£154	0	0
Expenses	78	0	0
	£78	0	0

Left to pay rent of cottage and garden, management, &c.; but it must be observed that returns begin to come in at the expiry of the first thirteen or fourteen weeks.

Estimates of the Cost, Expenses of Working, and Profits of a 5 Tray Patent Hydro-Incubator for one year, to produce 20 Fowls a day for market.

	£	s.	d.
Cost of incubator	105	0	0
2 Chicken mothers, at £15 15s. each, are requisite	31	10	0
9000 eggs for 18 broods in the year at 1d	37	10	0
75 chickens from every 100 eggs, gives 6750 chickens in the year, and the cost of their food up to 96 days old, will be 8d. each	225	0	0
Charcoal for the year	13	0	0
	£412	0	0

	£	s.	d.
6750 chickens, at 2s. each	675	0	0
Value of incubator and 2 chicken mothers	126	0	0
	801	0	0
Expenses	412	0	0
	£389	0	0

Left to pay rent of house, land, management, &c.

Estimate of the Cost, Expenses, and Profits of Working a 56-Tray Patent Hydro-Incubator, to produce 200 Fowls for Market every Day; with a View of the probable Expense of Erecting Premises, &c.; involving a Capital of £6,000, yielding upwards of 50 per cent. profit, after covering all Outlay.

Debit Account.
PREMISES.

	£	s.	d.
Cost of erecting a house for a Patent Hydro-Incubator, capable of hatching 300 eggs, to produce 200 fowls			

	£	s.	d.
a day (one-third being allowed for loss), 440 feet long by 24 feet wide, and 8 feet high on the ground floor, and 176 feet long on the second floor*	1200	0	0
Laying house for 1000 hens	125	0	0
Fatting do. 280 feet long by 16 feet wide	500	0	0
Fitting-up of roosting house and the laying house, with coops, &c. for the fatting house	250	0	0
Fencing	500	0	0
Fitting-up the offices and store-houses	400	0	0
Three roosting houses, with basking sheds, each containing 4 divisions for 2000 fowls, 60 feet long, 20 feet wide, and 8 feet high	375	0	0
	£3350	0	0

Such premises, along with 40 or 50 acres of poor sandy land, might be rented; and there are many places which could be for a small sum converted to answer the purpose.

Fixtures.

Licence for a 56-Tray Patent Hydro-Incubator, and 14 Chicken Mothers for rearing the poultry, including the cost of them at the manufactory	1550	0	0
Furniture of offices, 2 horses and 2 carts, harness, tools, troughs, &c.	300	0	0
Current expenses during the first quarter (before there are any returns), for labour, food of fowls, rent, taxes, &c., as per the following details	750	0	0
600 laying hens	90	0	0
	£6040	0	0

* The data for this is an estimate of a builder near London, who gave in £1 per foot, for building 12 feet wide, by 8 feet high.

	£.	s.	d
Rent of farm house and 40 or 50 acres of poor dry land, at £200 a year, one quarter	50	0	0
Superintendence, clerk, lighting, coals, taxes, and office expenses, £500 a year, one quarter	125	0	0
Labour, 12 men and 6 boys, market and other business expenses, feeding horses, &c., at the rate of £800 a year, one quarter	200	0	0
Feeding up 20,000 fowls, which is very nearly the number to which they will accumulate, of all ages, in 3 months, before any are fit for sale; and taking the cost of a full-grown fowl at 9d., the average of those on hand will be at the rate of 4½d. (8d. should pay for their food)	375	0	0
	£750	0	0

Credit Account.

	£.	s.	d
The produce of the year commencing after the first quarter will be 73,000 fowls fit for market, varying in value according to the present rate in London, from 2s. to 3s. each. Say the lowest price, 2s.	7200	0	0
The guano of about 20,000 fowls, which is the number alway on hand. I have estimated from 800 in my roosting house at Chiswick to be 56lb. a day from that number, or about 220 tons from the 20,000 fowls a year, say 80s. a ton.	880	0	0
The feathers of 73,000 fowls sold in the course of the year, I have ascertained from experience and calculation to yield 10,800lb. at 8d. . . .	360	0	0
Total annual revenue.	8440	0	0
Annual current expenses.	4237	10	0
	£4202	10	0

Annual Current Expenses.

	£.	s.	d.
Rent of farm house and land	200	0	0
Superintendence, Clerk, &c.	500	0	0
Labour, Market Expenses, &c.	800	0	0
Feeding 73,000 fowls a year at 9d.	2737	10	0
	£4237	10	0
Annual gross gain on a capital of £6000	4202	10	0

LIST OF PRICES

Of the CANTELONIAN PATENT HYDRO-INCUBATORS and Chicken Mothers, for Hatching Eggs and Rearing Poultry, with Estimates of the Cost, Produce, and Large Profits, to be gained by a Poultry Farm, Conducted on the Cantelonian Principle.

Orders for the Incubators, from 20 guineas each, and upwards, are solicited by SAMUEL GANT, the Manufacturer, 19, Tottenham Court New Road, and by Mr. CANTELO, at his temporary Model Farm, Chiswick.

The System is also in Operation at La Varenne de St. Maur, near Paris.

Agents throughout the Country will shortly be appointed.

EVERY one, who has room for it, ought to have a CANTELONIAN INCUBATOR, to rear poultry, and to produce new-laid eggs for breakfast, every morning. Noblemen and sportsmen are solicited to adopt these machines, to rear their game and fancy birds; and tradesmen, and public companies, to make profit. A Poultry yard attached to a Union Workhouse, would prove a great saving and advantage to feed the poor.

The smallest size Cantelonian Patent Hydro-Incubator is 4 feet long, 2 feet wide, and 2 feet 8 inches high; and is made to hatch 100 eggs. The price, cash, is 20 guineas.

The second size is for 200 eggs; 6 ft. 4 in. long, and the same height and width as the other. The price is 40 guineas, and so on to any extent that may be desired; 20 guineas being the price of every tray for 100 eggs—so that an Incubator made to hatch 1,000 chickens, would be 200 guineas, and so on in proportion.

The Portable Incubators are made for 100, 200, 400, and 600, but not larger. When premises are determined on, for a greater scale, it is better to fix the apparatus. *Printed instructions for their management will accompany the machines.*

The usual size of a Patent Chicken Mother is 44 feet long, and the price 15 guineas, cash, and so on, 15 guineas for every 44 feet long. These must always be fixed; and may be used to advantage, independently of the Incubators, to assist in rearing poultry by the ordinary slow method, and as a preventative against the roup, and other diseases, arising from cold and damp premises.

It is not pretended that the patent incubator will hatch and bring up every egg to a fowl. From 12 to 30 per cent., after great experience, has been found to be the discount. A one tray machine will enable the party who properly attends to it, to produce on the average 75 birds to a hatch, and 18 of these in the year, being 1,350 fowls. A very different result, indeed, to a hen, which sits but twice in the twelve months, and does not rear up above 8 chickens at a hatch. A two tray incubator and one mother will produce 2,700 a year, and so on in proportion—a thousand egg machine being capable of producing 13,500 full-grown fowls per annum. There is nothing in the principle to prevent millions of eggs being hatched 18 times in a year, by one machine.

Any person can, of course, calculate the sum they would amount to, according to the price at which he felt willing, or able, to sell his poultry.

Hens generally lay eggs after being six months old; but the Cantelonian system does not anticipate keeping a tenth of the poultry for laying stock, so that the quantity and profits arising from eggs, is not here taken into account.

Any cottager, or tradesman, who has a small garden, might attend to a one or two tray incubator in after-hours, and he may easily calculate what could be gained by a few thousand chickens.

To accommodate a five tray incubator, with two mothers, there ought to be a house or shed, 140 feet long by 12 feet wide, and 8 feet high,—or say 70 feet long by 24 feet wide, and an acre of ground for the poultry to exercise in, in the open air. The poorest land possible will do, provided it has a dry bottom.

It is only needful to multiply the above statements by any proportions desired, to give a fair result.

To estimate the value of the land and premises, to hatch chickens and rear poultry, is beyond the power of the Inventor of Practical Incubation by "*Top-contact Heat.*" An acre of land may be had for £10 or 10s. Sheds or poultry-houses may be erected to cost £10 or £100 a year. The principle is all that can be advanced regarding these matters, and he hopes the details have been clearly expressed, that the Cantelonian Patent Hydro Incubators will yield 18 hatches a year, and that the price is 20 guineas for every tray to hold 100 eggs.—Game, Turkies, Geese, Peacocks, Guinea Fowl, &c. can be produced on the same principle.

The price got for the birds, and the expense of management, must all depend on local circumstances.

Patents are secured for the United Kingdom and the Colonies, France, Belgium, Holland, and America; and the public must speedily be interested in the invention, as a new system of producing food. No party need to spend their money experimenting on the business. It has already, after enormous labour and cost, been brought to complete practical perfection, and now only requires to be prosecuted to gain wealth.

www.ingramcontent.com/pod-product-compliance
Lightning Source LLC
Chambersburg PA
CBHW062207220526
45470CB00009B/2954